Kindergarten Heart Unit Study

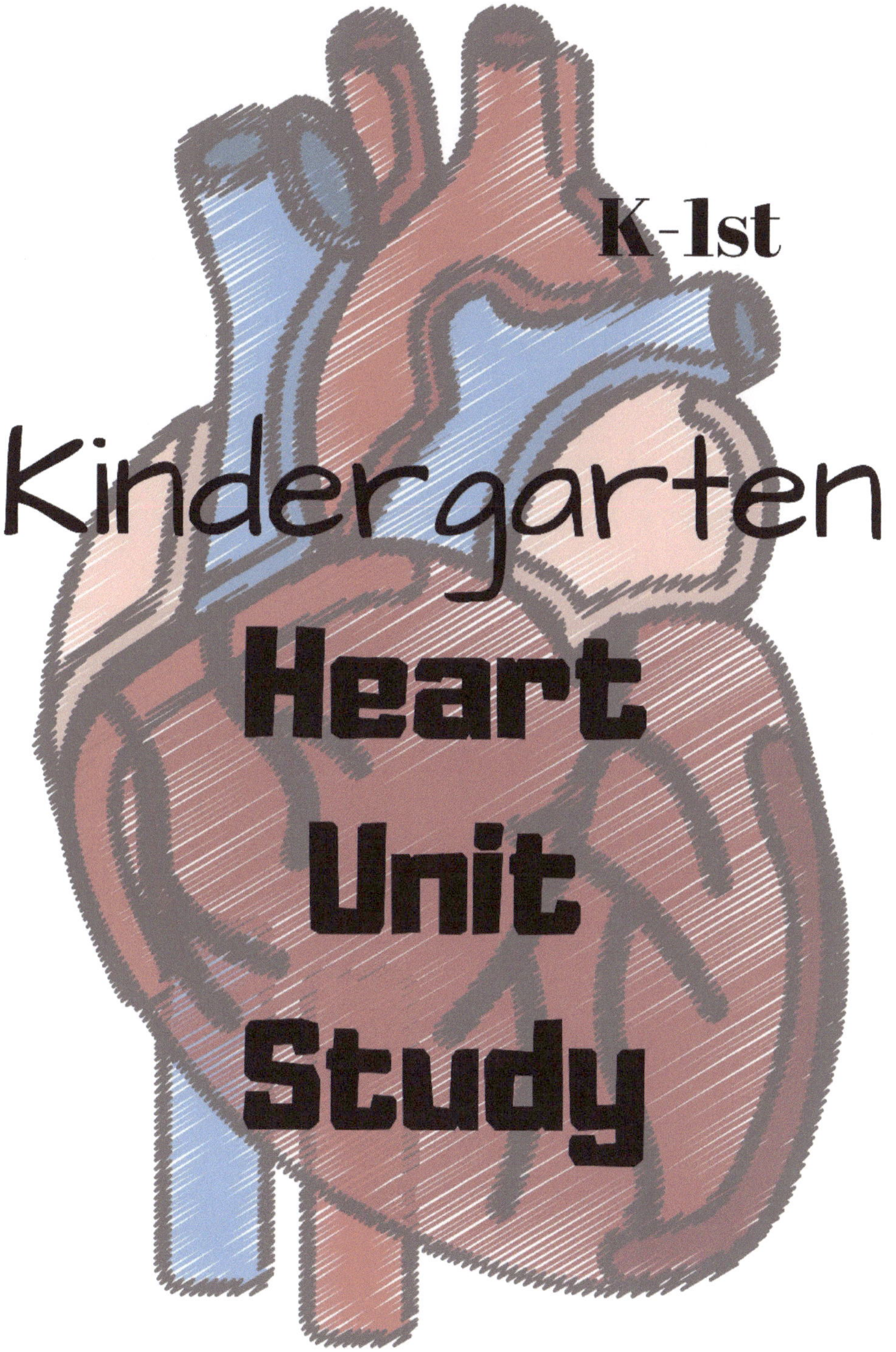

K-1st

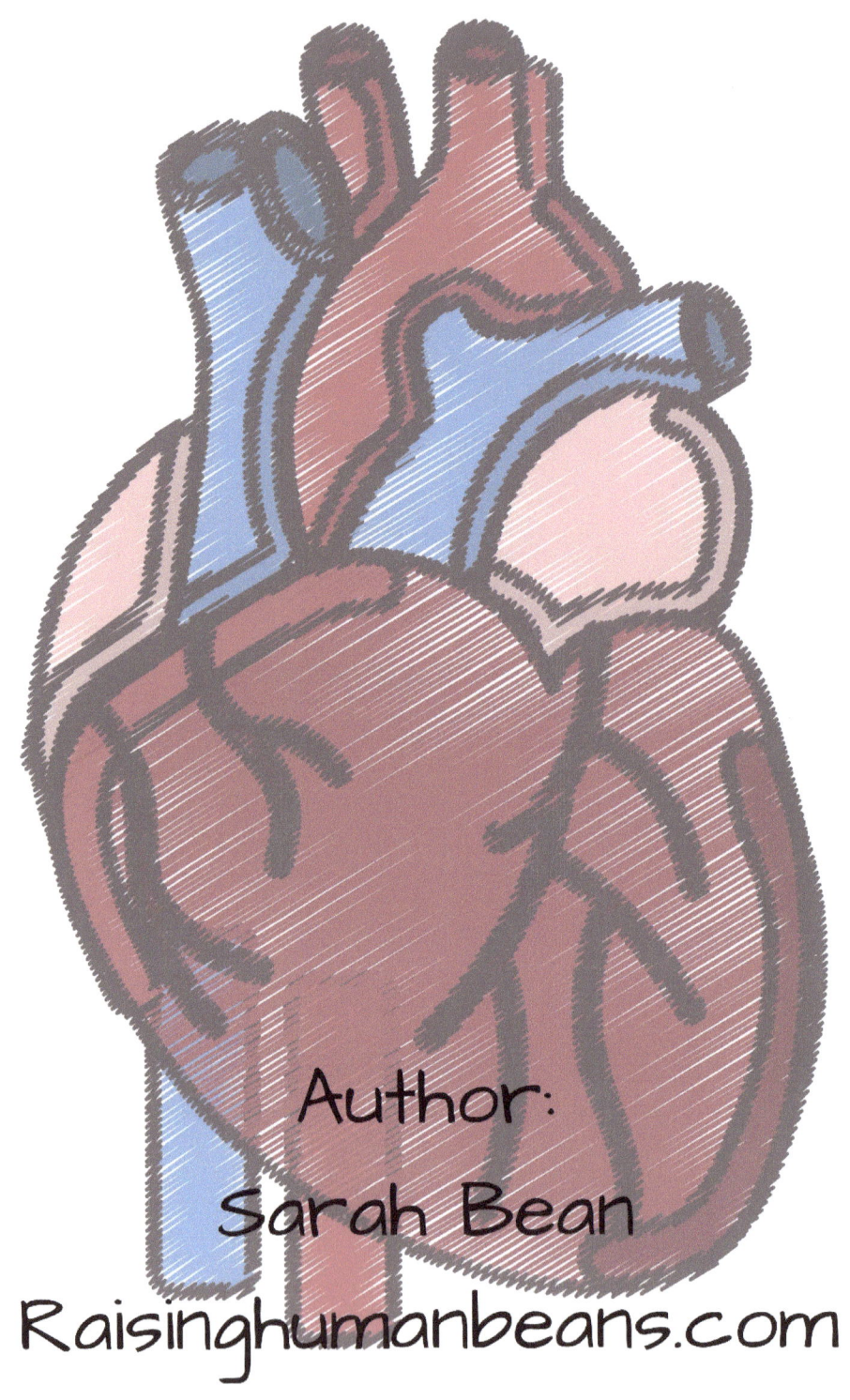

How to use this Study

In each Science study, there are 2-3 math pages; vocabulary, sight words, and a myriad of other activities and pages.

These Units were designed to take one week, in monthly conjunction with our State and President Studies; but you can extend the activities if you wish. I will soon have additional math lessons for purchase, and we are working on developing a Big Book of Unit Study, for K-6.

If you do follow the week approach, I recommend Math on MWF, Vocab and Sight Words Daily, and the other Activities Daily.

This is an open and go book, but planning for me always works better.

Good luck on your Homeschooling Journey!!

Note:

There are some pages that mention to cut out.

Please copy these pages if you wish to cut them out, so you still have the pages on the back.

Table of Contents

Page Number

1-2 Sight Words
3-5 Vocabulary
6 Facts
7-10 Parts of a Heart
11-12 How does it work?
13 Matching
14-17 Heart Math
18 Where is my heart?
19 Writing
20-21 Recipe

Heart Sight Words

Heart	Vein
Pump	Blood

Heart Sight Words

WEEK 1

For the first week, show these to your child every day. Make sure they are looking at the card, and go through all of them 3 times. Don't ask them to tell you what it is, just tell them what it is.

WEEK 2

1st round- Ask them what the words are. If they don't know, tell them.

2nd round- Put them on the ground in a circle. Ask them to jump to "this word." See if they are right!

3rd round- If they are still having trouble, explain to them why the word is the word. Have them remember based on sounding out the first few letters. Play the matching game a few times!

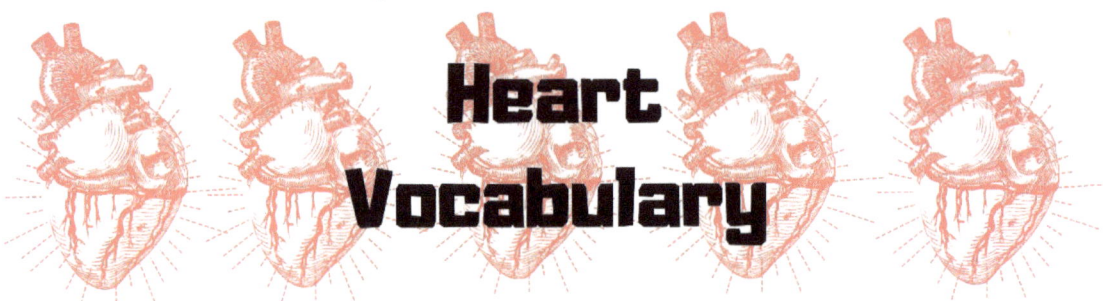

Heart Vocabulary

Vein:
A tube that carries blood with oxygen to the heart.

Artery:
A tube that carries blood without oxygen away from the heart to the lungs.

Muscle:
A band in the body that squeezes and becomes shorter (contracts) to move a part of the body.

Pump:
To force to move by moving in and out or up and down.

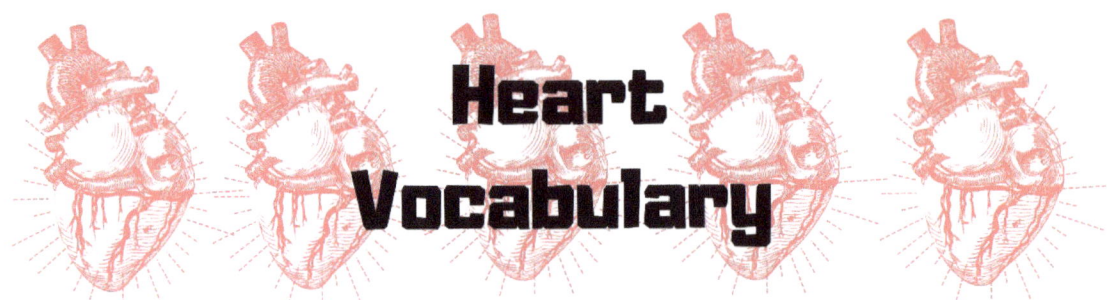

Heart Vocabulary

Practice these daily.

You can learn none of these; one of these; or all of these.

If you learn 1 or 2, this Unit could take 1 week.

If you learn 3-5, this Unit could take up to a month.

Remember: Go at your child's pace.

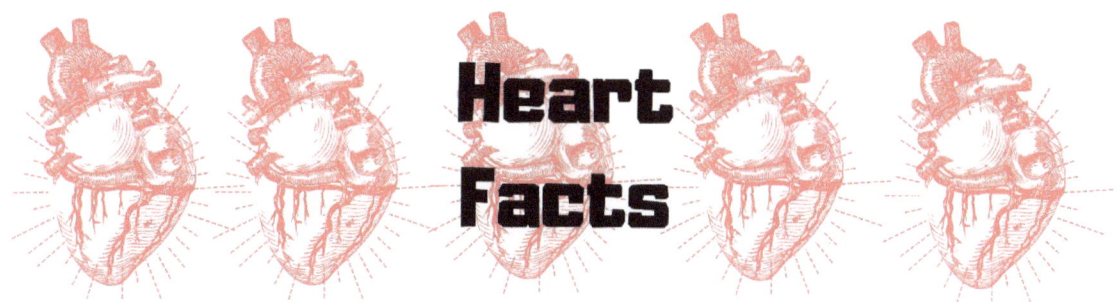

Heart Facts

1. A heart beats 100,000 times a day.
2. An adult heart pumps 5 quarts of blood a minute. (That's more than a gallon of milk)
3. If you attach all of your arteries, veins, and capillaries, it equals more than 60,000 miles. (That's if you went around the whole earth twice!)
4. Your heart's pressure can squirt up to 30 feet. (Measure 30 feet)
5. Your heart is the most important, and strongest muscle.
6. If you want to know how big your heart is, make a fist.
7. A pig and a human heart is the same size.

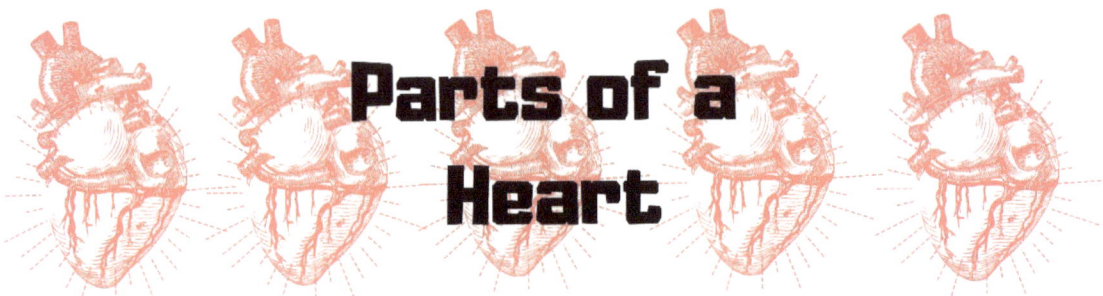

Parts of a Heart

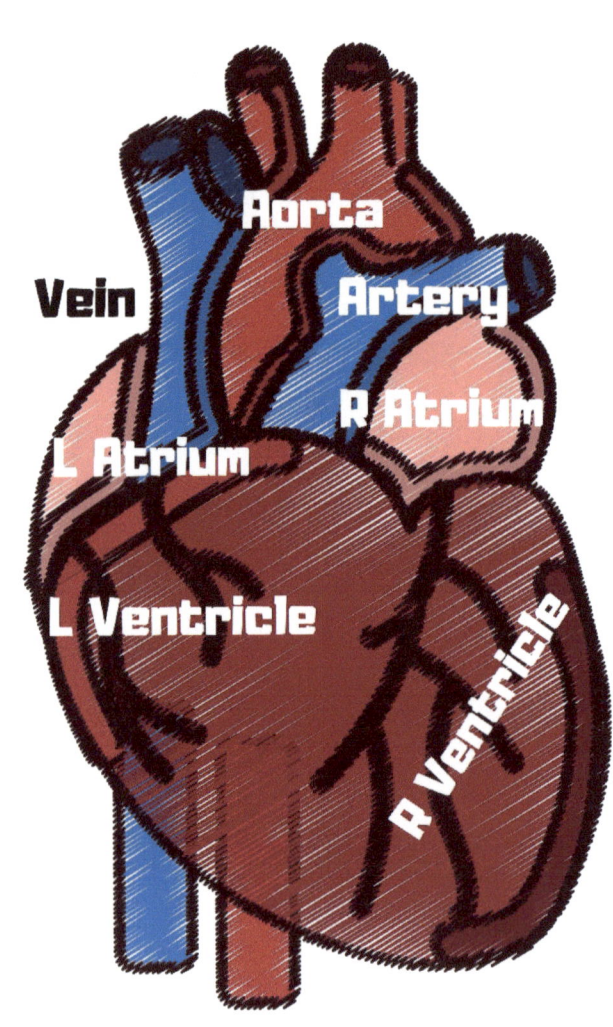

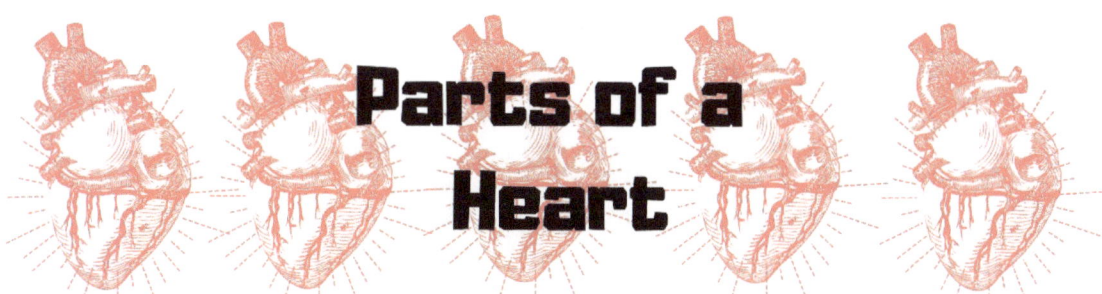
Parts of a Heart

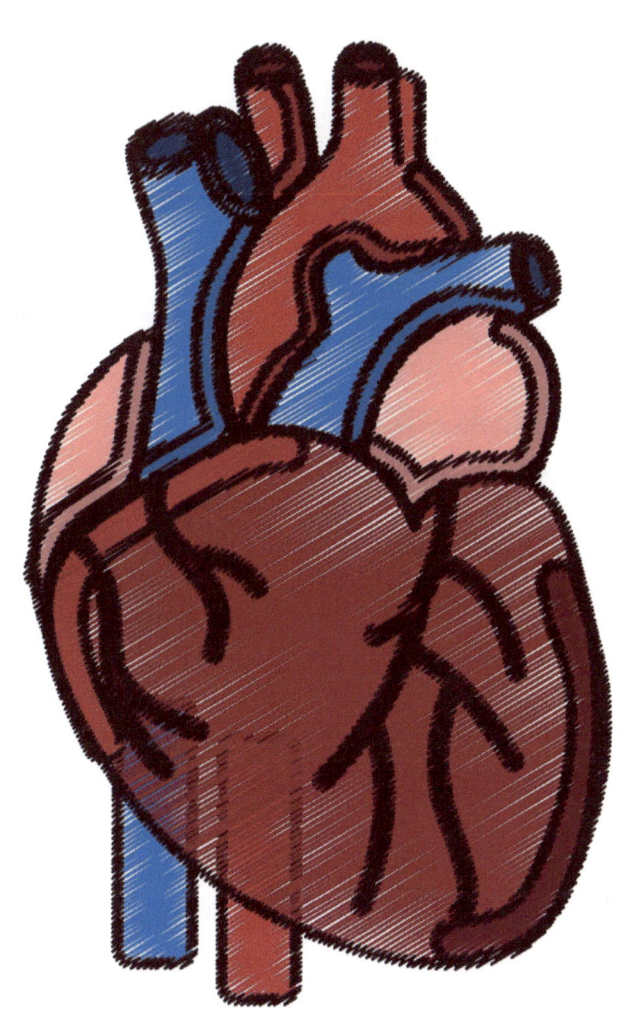

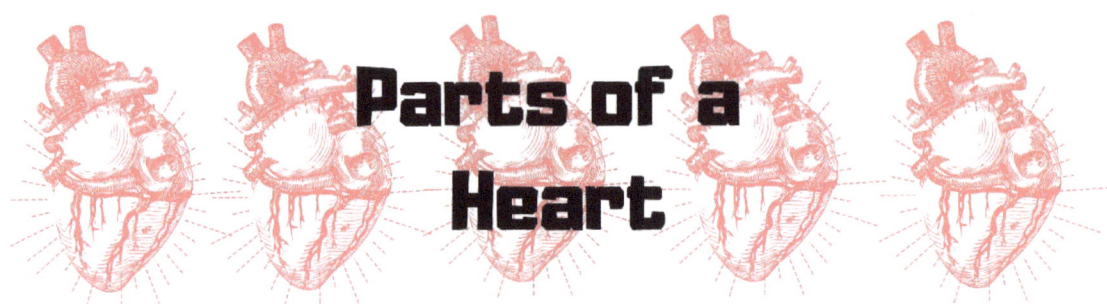

Cut and paste these onto the previous page of the heart.

- L Ventricle
- R Ventricle
- L Atrium
- R Atrium
- Aorta
- Vein
- Artery

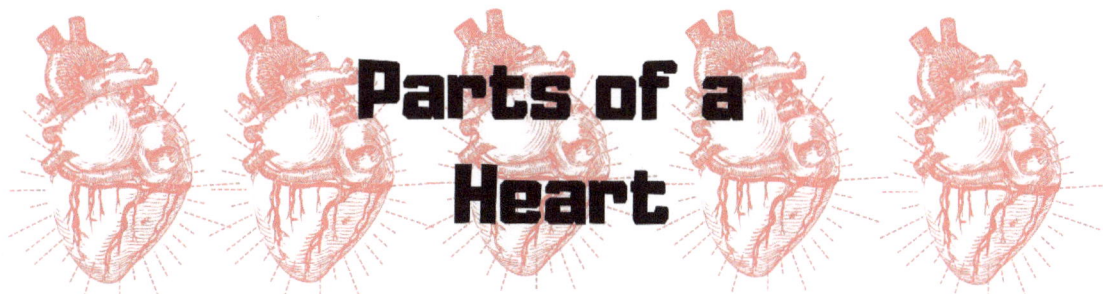

Parts of a Heart

Now using your heart picture above, please color the picture using these instructions.

Color the Aorta red, The Ventricles Orange, The Atriums pink, and the Arteries Blue.
Draw a Red Vein coming out of the left side of the artery.

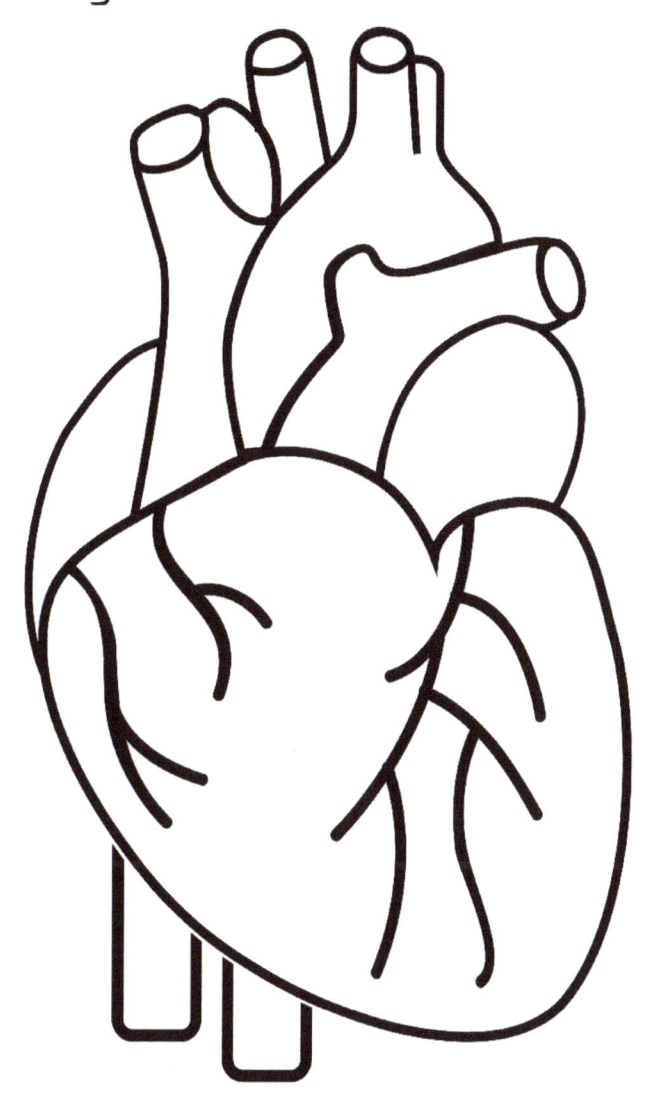

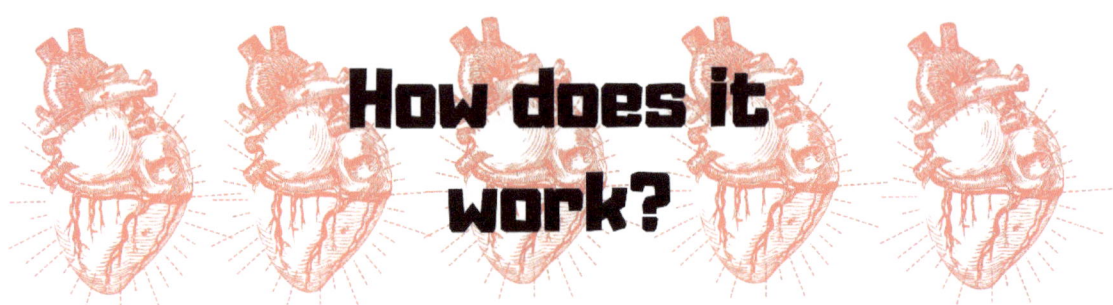

How does it work?

A heart has a Pump Cycle. This is how it goes.

1. Veins bring blood with oxygen to the heart.
2. Blood enters the right atrium, and then the right ventricle.
3. Blood exits the Right ventricle and travels into the Arteries.
4. The arteries carry it to the lungs, where it receives oxygen.
5. The blood goes through the veins into the Left Atrium, then into the Left Ventricle.
6. From the Left Ventricle, through the Aorta.
7. Then back to the veins.

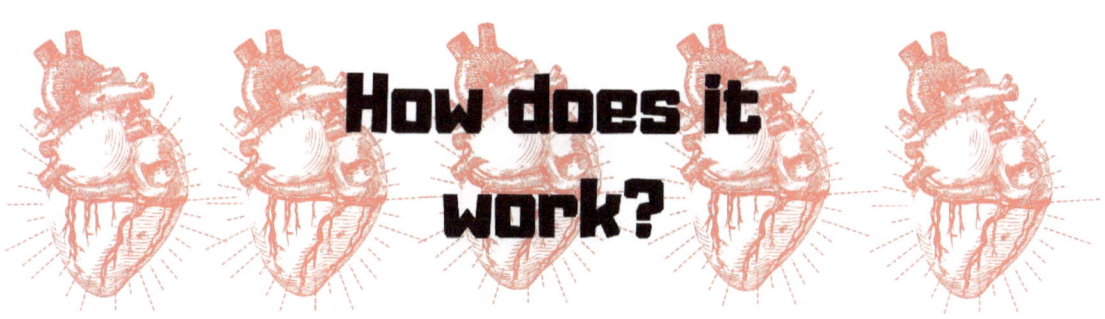

How does it work?

Memorize the pump cycle.

1. Veins
2. Right Atrium
3. Right Ventricle
4. Arteries
5. Lungs
6. Oxygen
7. Veins
8. Left Atrium
9. Left Ventricle
10. Aorta
11. Veins
12. Right Atrium

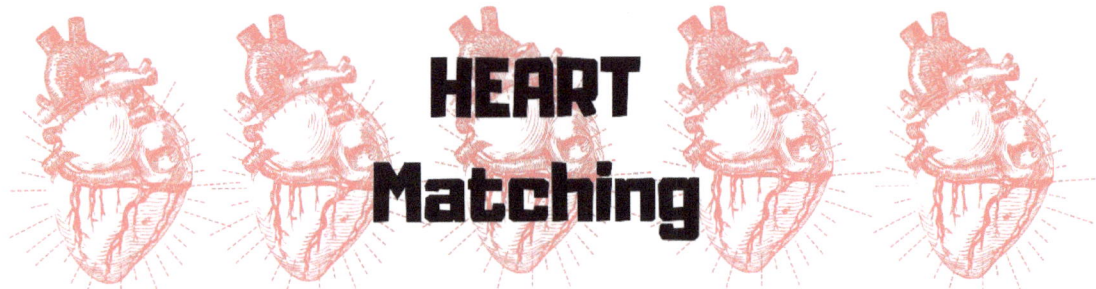

HEART Matching

Heart

Vein

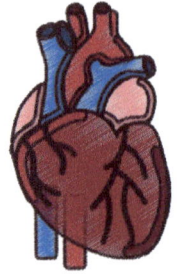

Pump

Blood

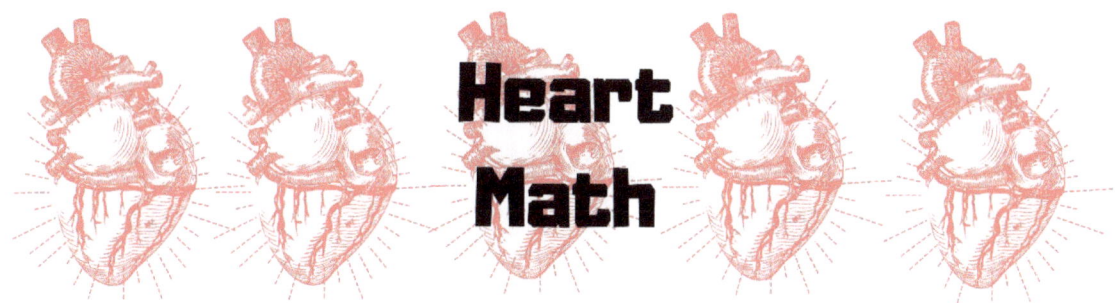

Heart Math

Cut out the shapes, and sort them on the next page.

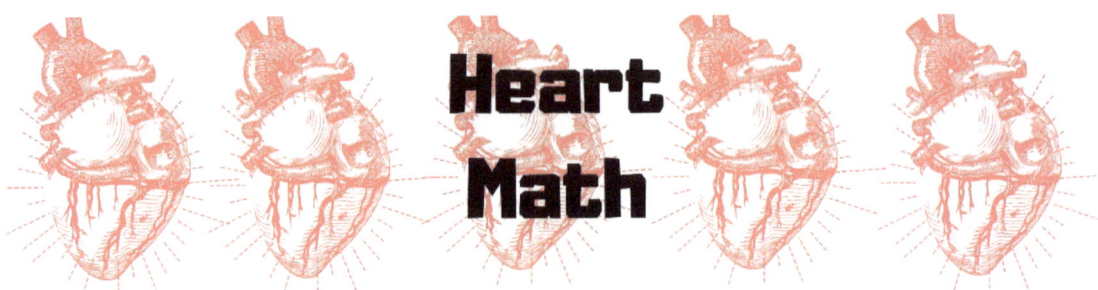

Heart Math

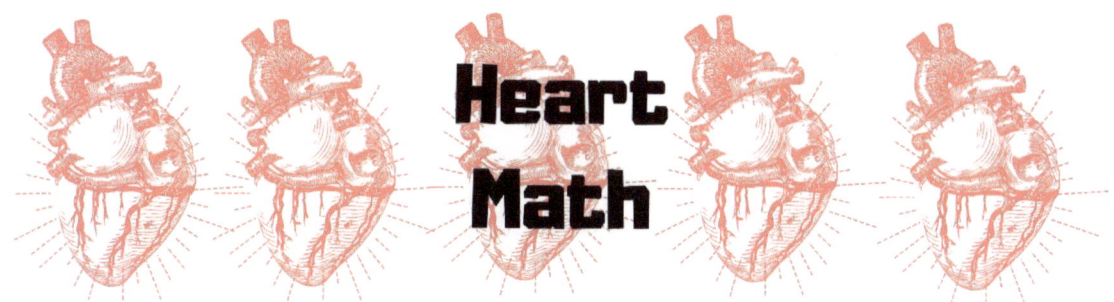

Circle Which shape has the most?

Circle Which shape has the least?

How many of each shape is there?

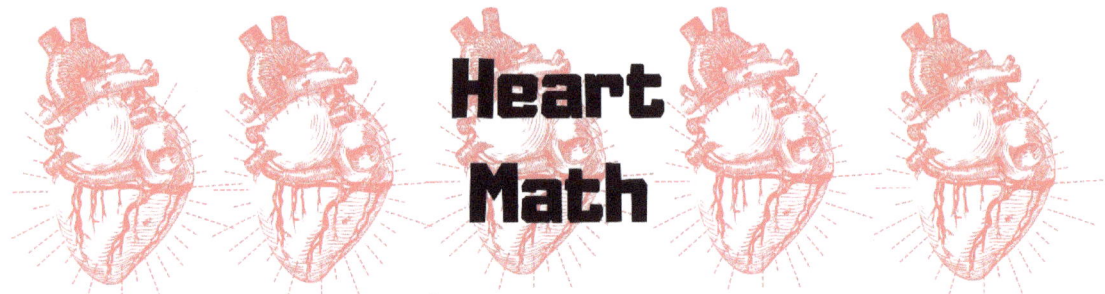

Heart Math

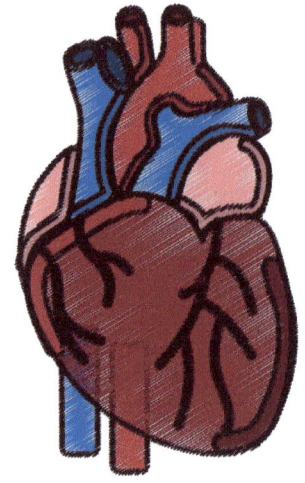 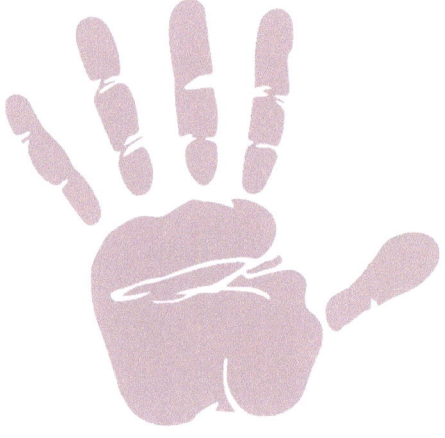

Measure this heart, how many inches is it?

Measure your hand, in a fist. How many inches is it?

_____   _____

Is this heart, or your hand, bigger?

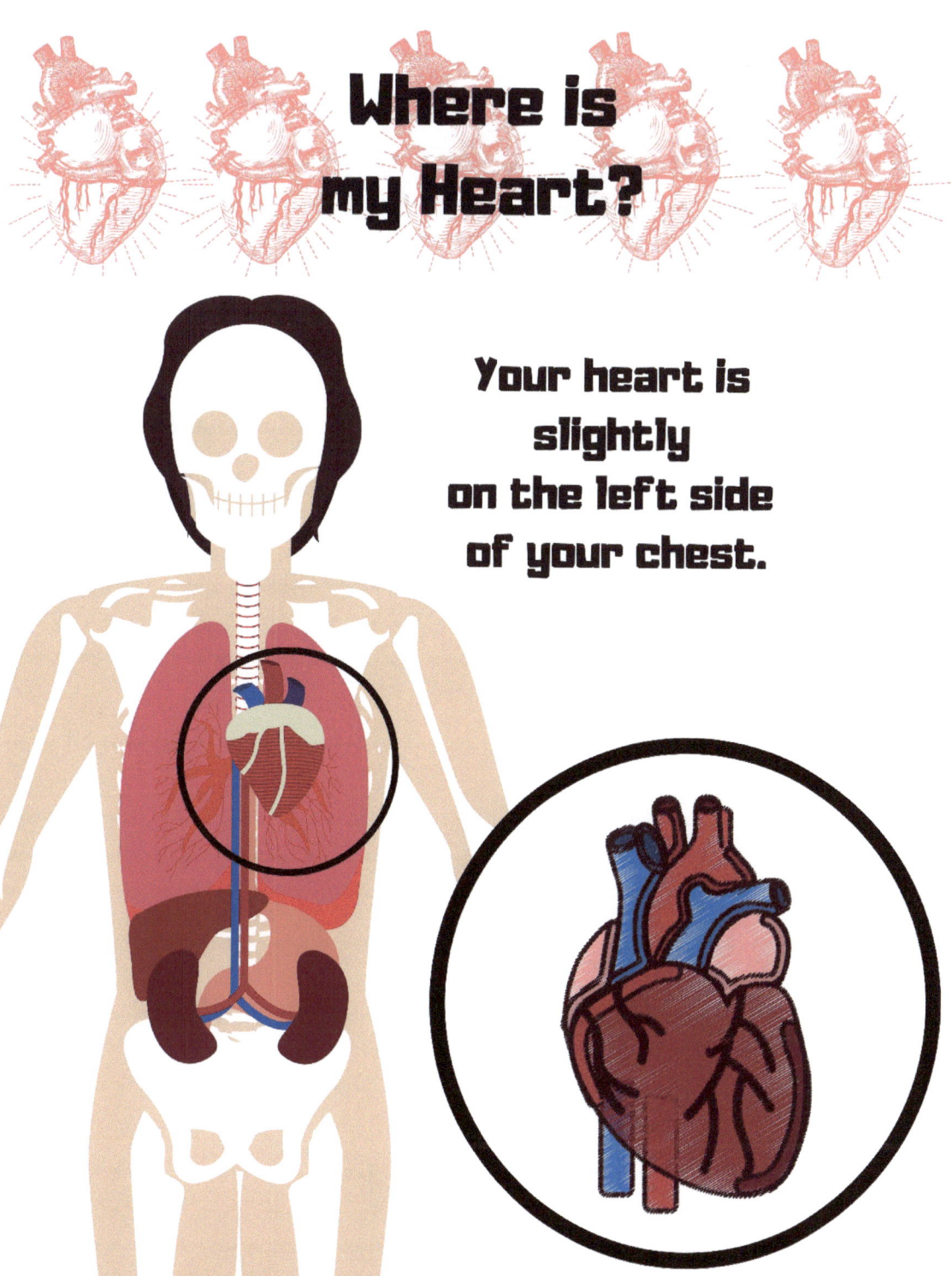

Heart Writing

Please copy these words.

California

Heart

Heart Shaped Raspberry Rolls

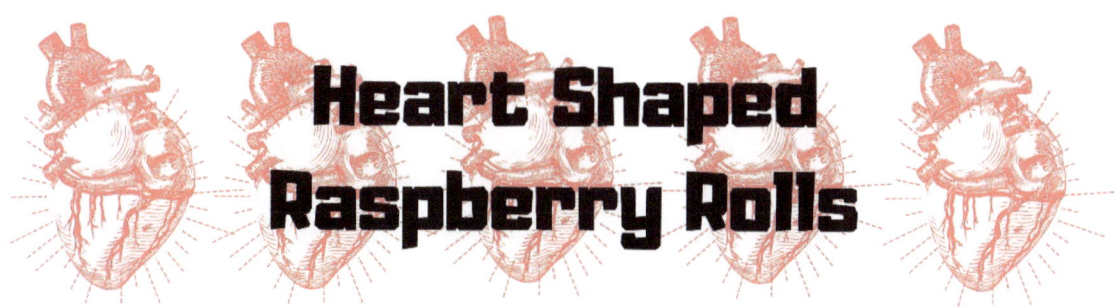

Ingredients:

For the dough:
1 cup Raspberry Greek yogurt
1.5 Cups flour
1.5 tsp baking powder
.25 tsp salt

For the filling:
16oz Cream Cheese
3/4 Cup Sugar

Heart Shaped Raspberry Rolls

Directions:
1. Make the dough by mixing the dry ingredients, then add the yogurt.
2. Mix thoroughly.
3. Roll it out, into an oval.
4. Mix the cream cheese and sugar until thoroughly mixed. Taste, add sugar as needed.
5. Spread cream cheese on dough. Roll each side halfway until they meet in the middle. Cut and sqeeze together to make a heart shape.
6. Bake at 450* for 10-15 minutes.
7. Drizzle with white chocolate, or cream cheese

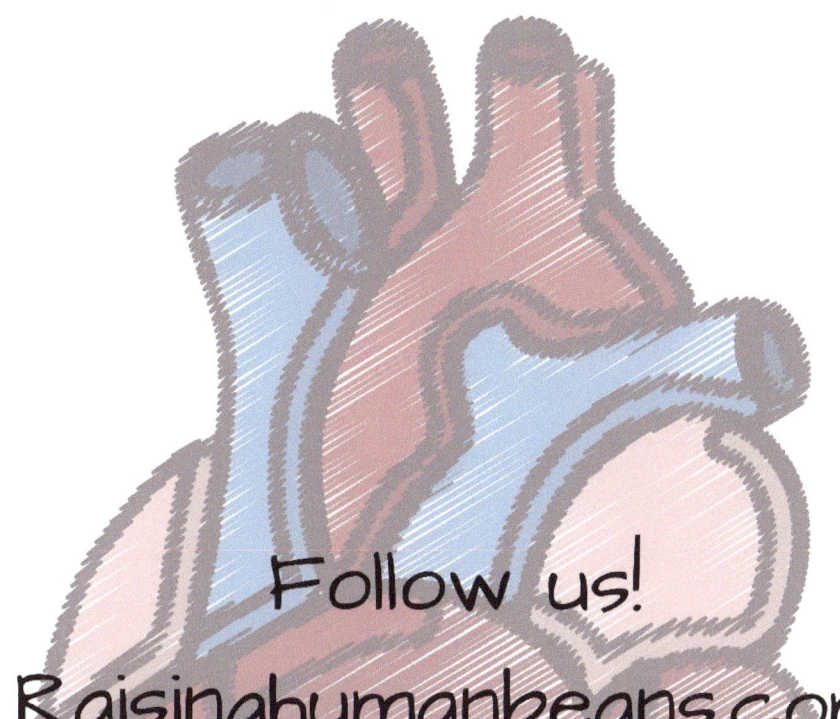

Follow us!
Raisinghumanbeans.com
Facebook.com/raisingbeans
Instagram.com/raisinghumanbeans
Pinterest.com/raisinghumanbeans

Raising Human Beans
2018

www.ingramcontent.com/pod-product-compliance
Lightning Source LLC
Chambersburg PA
CBHW042322250526
R18347300001B/R183473PG45473CBX00013B/1